Weekly Schedule

Time / Period	Monday	Tuesday	Wednesday	Thursday	Friday	Saturday	Sunday

AF887863

Emergency Contacts

Name:	Relation:	Cell Phone:	Phone:	Email:
Address:		City:	State:	Zip Code:

Name:	Relation:	Cell Phone:	Phone:	Email:
Address:		City:	State:	Zip Code:

Academic Calendar

Year:

SPRING SEMESTER 2

Day														

January

1	2	3	4	5	6	7	8	9	10	11	12	13	14
15	16	17	18	19	20	21	22	23	24	25	26	27	28
29	30	31											

February

			1	2	3	4	5	6	7	8	9	10	11
12	13	14	15	16	17	18	19	20	21	22	23	24	25
26	27	28	29										

Change day if it is a leap year (2024, 2028, 2032, 2036, 2040, 2044, 2048, 2052, 2056, 2060. . .)

Day														

March

			1	2	3	4	5	6	7	8	9	10	11
12	13	14	15	16	17	18	19	20	21	22	23	24	25
26	27	28	29	30	31								

April

						1	2	3	4	5	6	7	8
9	10	11	12	13	14	15	16	17	18	19	20	21	22
23	24	25	26	27	28	29	30						

SUMMER SEMESTER 3

May

	1	2	3	4	5	6	7	8	9	10	11	12	13
14	15	16	17	18	19	20	21	22	23	24	25	26	27
28	29	30	31										

June

				1	2	3	4	5	6	7	8	9	10
11	12	13	14	15	16	17	18	19	20	21	22	23	24
25	26	27	28	29	30								

July

						1	2	3	4	5	6	7	8
9	10	11	12	13	14	15	16	17	18	19	20	21	22
23	24	25	26	27	28	29	30	31					

FALL SEMESTER 1

August

		1	2	3	4	5	6	7	8	9	10	11	12
13	14	15	16	17	18	19	20	21	22	23	24	25	26
27	28	29	30	31									

September

					1	2	3	4	5	6	7	8	9
10	11	12	13	14	15	16	17	18	19	20	21	22	23
24	25	26	27	28	29	30							

October

1	2	3	4	5	6	7	8	9	10	11	12	13	14
15	16	17	18	19	20	21	22	23	24	25	26	27	28
29	30	31											

November

			1	2	3	4	5	6	7	8	9	10	11
12	13	14	15	16	17	18	19	20	21	22	23	24	25
26	27	28	29	30									

December

					1	2	3	4	5	6	7	8	9
10	11	12	13	14	15	16	17	18	19	20	21	22	23
24	25	26	27	28	29	30	31						

Events and Due Dates

Title:	Start:	End:
Note:		

Title:	Start:	End:
Note:		

Title:	Start:	End:
Note:		

Title:	Start:	End:
Note:		

Title:	Start:	End:
Note:		

Title:	Start:	End:
Note:		

Assignment / Project / Exam	Due Date

Assignment / Project / Exam	Due Date

Copyright & Disclaimer

NOTABLE+ MATH NOTEBOOK: G3E, COOL DRIP, STANDARD+E (120 PAGES)

Published by Notable Plus LLC.

The author is represented by Notable Plus LLC, Palm Bay, Florida.

Visit our website at www.notableplus.com

Notable+ Math Notebook: G3E, Cool Drip, Standard+E (120 Pages)
Paperback: 120 Pages
Print Quality: Black ink and 55# (90 GSM) white paper
ISBN-10: 1-960176-11-0
ISBN-13: 978-1-960176-11-0

Notable+ also publishes it books and content in a variety of electronic formats and online. Some of the content that appears in print may not be available in electronic books or online and some of the content that appears online or in electronic formats may not be available in print.

Contents (Fillable)

Notables Contents

My Notables

Date: / /

Date: / /

Date: / /

Date: / /

Date: / /

Date: / /

Date: / /

Date: / /

Date: / /

Date: / /

Date: / /

Date: / /

Date: / /

Date: / /

Date: / /

Date: / /

Date: / /

Date: / /

Date: / /

Date: / /

Date: / /

Date: / /

Date: / /

Date: / /

Date: / /

Date: / /

Date: / /

Date: / /

Date: / /

Date: / /

Date: / /

Date: / /

Date: / /

Date: / /

Date: / /

Date: / /

Date: / /

Date: / /

Date: / /

Date: / /

Date: / /

Date: / /

Date: / /

Date: / /

Date: / /

Date: / /

Date: / /

Date: / /

Date: / /

Date: / /

Date: / /

Date: / /

Date: / /

Date: / /

Date: / /

Date: / /

Date: / /

Date: / /

Date: / /

Date: / /

Date: / /

Date: / /

Date: / /

Date: / /

Date: / /

Date: / /

Math Notables

1 Number Place Value Chart

Digits	Places	
1	Ones	1
2	Tens	10
3	Hundreds	100
4	Thousands	1,000
5	Ten Thousands	10,000
6	Hundred Thousands	100,000
7	Millions	1,000,000
8	Ten Millions	10,000,000
9	Hundred Millions	100,000,000
10	Billions	1,000,000,000

Digits	Decimals	Places	Fractions
	.	Decimal Point	
1	.1	Tenths	1/10
2	.01	Hundredths	1/100
3	.001	Thousandths	1/1,000
4	.0001	Ten Thousandths	1/10,000
5	.00001	Hundred Thousandths	1/100,000
6	.000001	Millionths	1/1,000,000
7	.0000001	Ten Millionths	1/10,000,000
8	.00000001	Hundred Millionths	1/100,000,000
9	.000000001	Billionths	1/1,000,000,000

2 Multiplication Tables

1 x 0 = 0	2 x 0 = 0	3 x 0 = 0	4 x 0 = 0	5 x 0 = 0	6 x 0 = 0
1 x 1 = 1	2 x 1 = 2	3 x 1 = 3	4 x 1 = 4	5 x 1 = 5	6 x 1 = 6
1 x 2 = 2	2 x 2 = 4	3 x 2 = 6	4 x 2 = 8	5 x 2 = 10	6 x 2 = 12
1 x 3 = 3	2 x 3 = 6	3 x 3 = 9	4 x 3 = 12	5 x 3 = 15	6 x 3 = 18
1 x 4 = 4	2 x 4 = 8	3 x 4 = 12	4 x 4 = 16	5 x 4 = 20	6 x 4 = 24
1 x 5 = 5	2 x 5 = 10	3 x 5 = 15	4 x 5 = 20	5 x 5 = 25	6 x 5 = 30
1 x 6 = 6	2 x 6 = 12	3 x 6 = 18	4 x 6 = 24	5 x 6 = 30	6 x 6 = 36
1 x 7 = 7	2 x 7 = 14	3 x 7 = 21	4 x 7 = 28	5 x 7 = 35	6 x 7 = 42
1 x 8 = 8	2 x 8 = 16	3 x 8 = 24	4 x 8 = 32	5 x 8 = 40	6 x 8 = 48
1 x 9 = 9	2 x 9 = 18	3 x 9 = 27	4 x 9 = 36	5 x 9 = 45	6 x 9 = 54
1 x 10 = 10	2 x 10 = 20	3 x 10 = 30	4 x 10 = 40	5 x 10 = 50	6 x 10 = 60
1 x 11 = 11	2 x 11 = 22	3 x 11 = 33	4 x 11 = 44	5 x 11 = 55	6 x 11 = 66
1 x 12 = 12	2 x 12 = 24	3 x 12 = 36	4 x 12 = 48	5 x 12 = 60	6 x 12 = 72

7 x 0 = 0	8 x 0 = 0	9 x 0 = 0	10 x 0 = 0	11 x 0 = 0	12 x 0 = 0
7 x 1 = 7	8 x 1 = 8	9 x 1 = 9	10 x 1 = 10	11 x 1 = 11	12 x 1 = 12
7 x 2 = 14	8 x 2 = 16	9 x 2 = 18	10 x 2 = 20	11 x 2 = 22	12 x 2 = 24
7 x 3 = 21	8 x 3 = 24	9 x 3 = 27	10 x 3 = 30	11 x 3 = 33	12 x 3 = 36
7 x 4 = 28	8 x 4 = 32	9 x 4 = 36	10 x 4 = 40	11 x 4 = 44	12 x 4 = 48
7 x 5 = 35	8 x 5 = 40	9 x 5 = 45	10 x 5 = 50	11 x 5 = 55	12 x 5 = 60
7 x 6 = 42	8 x 6 = 48	9 x 6 = 54	10 x 6 = 60	11 x 6 = 66	12 x 6 = 72
7 x 7 = 49	8 x 7 = 56	9 x 7 = 63	10 x 7 = 70	11 x 7 = 77	12 x 7 = 84
7 x 8 = 56	8 x 8 = 64	9 x 8 = 72	10 x 8 = 80	11 x 8 = 88	12 x 8 = 96
7 x 9 = 63	8 x 9 = 72	9 x 9 = 81	10 x 9 = 90	11 x 9 = 99	12 x 9 = 108
7 x 10 = 70	8 x 10 = 80	9 x 10 = 90	10 x 10=100	11 x 10=110	12 x 10=120
7 x 11 = 77	8 x 11 = 88	9 x 11 = 99	10 x 11=110	11 x 11=121	12 x 11=132
7 x 12 = 84	8 x 12 = 96	9 x 12=108	10 x 12=120	11 x 12=132	12 x 12=144

Metric & Imperial Conversion Tables

3 Length

Metric Unit	Base Unit Equivalent	Imperial / US Equivalent
micrometer (μm)	0.000001 meter	0.000039 inch
millimeter (mm)	0.001 meter	0.039 inch
centimeter (cm)	0.01 meter	0.39 inch
decimeter (dm)	0.1 meter	3.94 inches
meter (m)		39.37 inches
dekameter (dkm)	10 meters	32.81 feet
hectometer (hm)	100 meters	328.08 feet
kilometer (km)	1,000 meters	0.62 mile

4 Capacity

Metric Unit	Base Unit Equivalent	Imperial / US Equivalent	
		(cubic)	(liquid)
microliter (μl)	0.000001 liter	0.00027 fluid dram	0.000061 cubic inch
milliliter (ml)	0.001 liter	0.27 fluid dram	0.061 cubic inch
centiliter (cl)	0.01 liter	0.34 fluid ounce	0.61 cubic inch
deciliter (dl)	0.1 liter	0.21 pint	6.10 cubic inches
liter (l)		1.06 quarts	61.02 cubic inches
dekaliter (dkl)	10 liters	2.64 gallons	0.35 cubic foot
hectoliter (hl)	100 liters	26.42 gallons	3.53 cubic feet
Kiloliter (kl)	1,000 liters	264.17 gallons	1.31 cubic yards

Metric & Imperial Conversion Tables

5 Weight

Metric Unit	Base Unit Equivalent	Empirical / US Equivalent
microgram (µg)	0.000001 gram	0.000015 grain
milligram (mg)	0.001 gram	0.015 grain
centigram (cg)	0.01 gram	0.15 grain
decigram (dg)	0.1 gram	1.54 grain
gram (g)		0.035 ounce
dekagram (dkg)	10 grams	0.35 ounces
hectogram (hg)	100 grams	3.53 ounces
kilogram (kg)	1000 grams	2.20 pounds
metric ton (t)	1,000,000 grams	1.10 short tons

6 Area

Metric Unit	Base Unit Equivalent	Empirical / US Equivalent
square centimeter (sq cm or cm^2)	0.0001 square meter	0.16 square inch
are (a)	100 square meters	119.60 square yards
hectare (ha)	10,000 square meters	2.47 acres
square kilometer (sq km or km^2)	1,000,000 square meters	0.39 square mile

7 Volume

Metric Unit	Base Unit Equivalent	Empirical / US Equivalent
cubic centimeter (cu cm, cm^3 or cc)	0.000001 cubic meter	0.061 cubic inch
cubic decimeter (dm^3)	0.001 cubic meter	61.02 cubic inches
cubic meter (m^3)		1.31 cubic yards

8 Temperature Conversion

Celsius to Fahrenheit	Fahrenheit to Celsius	Celsius to Kelvin
(°C × 9/5) + 32 = °F	(°F – 32) × 5/9 = °C	°C + 273.15 = K

9 Area (A) of Two-Dimensional Figures

Rectangle	$A = lw$, where l is the length and w is the width $A = bh$, where b is the base and h is the height
Square	$A = lw$, where l is the length and w is the width $A = bh$, where b is the base and h is the height $A = s^2$, where s is the side length
Triangle	$A = \frac{1}{2} bh$, where b is the base and h is the height
Circle	$A = \pi r^2$, where r is the radius
Trapezoid	$A = \frac{1}{2} (b_1 + b_2)h$, where b_1 and b_2 are the bases and h is the height
Parallelogram	$A = bh$, where b is the base and h is the height
Rhombus	$A = bh$, where b is the base and h is the height $A = \frac{1}{2} d_1 d_2$, where d_1 and d_2 are the diagonals
Equilateral Triangle	$A = \sqrt{3}/4s^2$, where s is the side length
Regular Polygon	$A = \frac{1}{2} Pa$, where P is the perimeter and a is the apothem

Order of Operations

10 **PEMDAS** Parentheses, Exponents, Multiplication, Division, Addition, Subtraction

11 **BODMAS** Brackets, Order or Off, Parentheses, Exponents, Multiplication, Division, Addition, Subtraction

12 Surface Area (SA) of Three-Dimensional Figures

Cube	$SA = 6s^2$, where s is the side length
Prism	$SA = 2B + Ph$, where B is the area of the base, P is the perimeter of the base and h is the height
Cylinder	$SA = 2B + Ph$, where B is the area of the base, P is the perimeter of the base and h is the height
Cone	$SA = B + \pi r h_s$, where B is the area of the base, r is the radius and h_s is the slant height
Pyramid	$SA = B + A(each\ face)$, where B is the area of the base and A(each face) is the area of each face
Regular pyramid	$SA = B + \frac{1}{2} Ph_s$, where B is the area of the base, P is the perimeter of the base and h_s is the slant height
Sphere	$SA = 4\pi r^2$, where r is the radius
Hemisphere	$SA = 3\pi r^2$, where r is the radius and the area of the flat side is included

13 Volume (V) of Three-Dimensional Figures

Cube	$V = s^3$, where s is the side length
Prism	$V = Bh$, where B is the area of the base and h is the height
Cylinder	$V = Bh$, where B is the area of the base and h is the height
Cone	$V = \frac{1}{3}Bh$, where B is the area of the base and h is the height
Pyramid	$V = \frac{1}{3}Bh$, where B is the area of the base and h is the height
Sphere	$V = 4/3\pi r^3$, where r is the radius

14 Laws of Exponents

(where m and n are integers and a and b are nonzero real numbers)

Product of powers	$a^m * a^n = a^{m+n}$ and conversely $a^{m+n} = a^m * a^n$
Quotient of powers	$\frac{a^m}{a^n} = a^{m-n}$ and conversely $a^{m-n} = \frac{a^m}{a^n}$
Power of a power	$(a^m)^n = a^{m*n}$ and conversely $a^{m*n} = (a^m)^n$
Power of a product	$(ab)^m = a^m * b^m$ and conversely $a^m * b^m = (ab)^m$
Power of a quotient	$(\frac{a}{b})^m = \frac{a^m}{b^m}$ and conversely $\frac{a^m}{b^m} = (\frac{a}{b})^m$
Negative exponent	$a^{-1} = \frac{1}{a}$ and conversely $\frac{1}{a} = a^{-1}$ $(\frac{a}{b})^{-1} = \frac{b}{a}$ and conversely $\frac{b}{a} = (\frac{a}{b})^{-1}$
Identity exponent	$a^1 = a$
Zero exponent	$a^0 = 1$
Rational, Fractional exponent	$a^{\frac{m}{n}} = (\sqrt[n]{a})^m$ and conversely $(\sqrt[n]{a})^m = a^{\frac{m}{n}}$, where $a > 0$ $a^{\frac{m}{n}} = \sqrt[n]{(a^m)}$ and conversely $\sqrt[n]{(a^m)} = a^{\frac{m}{n}}$, where $a > 0$

Mean, Median, Mode and Range

(Mean, median, and mode are three kinds of "averages".)

15 **Mean** is found by adding the numbers and dividing the sum by the number of numbers in the list.

16 **Median** is the "middle" value in the list of numbers ordered from smallest to largest.

17 **Mode** is the value that occurs most often. If no number in the list is repeated, then there is no mode for the list.

18 **Range** is the difference between the largest and smallest values of a list of numbers. It expresses "spread" (or how distributed or concentrated the values are).

Basic Finance

(Level 1)

19 **Discount, Selling Price and Listed Price**

20 **Discount** a deduction from the usual cost of something

21 **Listed Price** or MRP (maximum retail price) of the product is the actual price of an article as declared by the manufacturer. Listed Price is the price before the discount is applied.

22 **Selling Price** is the price after the discount is applied

Discount = Listed Price - Selling Price
Selling Price = Listed Price – Discount
Listed Price = Selling Price + Discount

If a discount is expressed as a percentage (rate) then the discount formula is:
Discount = Listed Price × Discount Rate
Rate of Discount = Discount% = (Discount/Listed Price) ×100
Selling Price = Listed Price [(100–discount%)/100]
Listed Price = (Selling Price × 100)/(100–discount%)

23 **Simple Interest** = P × R × T

24 **Principal (P)** is the amount that initially borrowed from the bank or invested.

25 **Rate (R)** is the rate of interest at which the principal amount is given to someone for a certain time, the rate of interest can be 5%, 10%, or 15%, etc.

26 **Time (T)** is the duration for which the principal amount is given to someone.

27 **Amount (A)** When a person takes a loan from a bank, he/she has to return the principal borrowed plus the interest amount, and this total returned is called Amount.

Basic Trigonometric Functions

(Solving Problems with Triangles)

Pythagorean Theorem

28 **Pythagorean Theorem** $a^2 + b^2 = c^2$

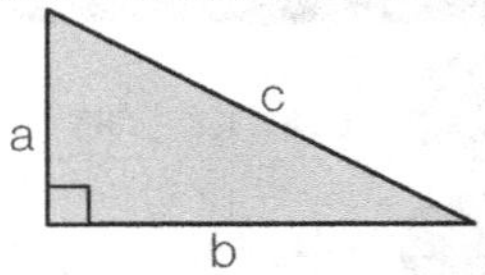

a = side of right triangle

b = side of right triangle

c = hypotenuse (The hypotenuse is the longest side of a right-angled triangle, the side opposite the right angle.)

Sine Cosine and Tangent for Right-Angled Triangles

"Opposite" is opposite to the angle θ

"Adjacent" is adjacent (next to) to the angle θ

"Hypotenuse" is the long one

29 **Sine** $\sin\theta = \dfrac{\text{Opposite}}{\text{Hypotenuse}}$

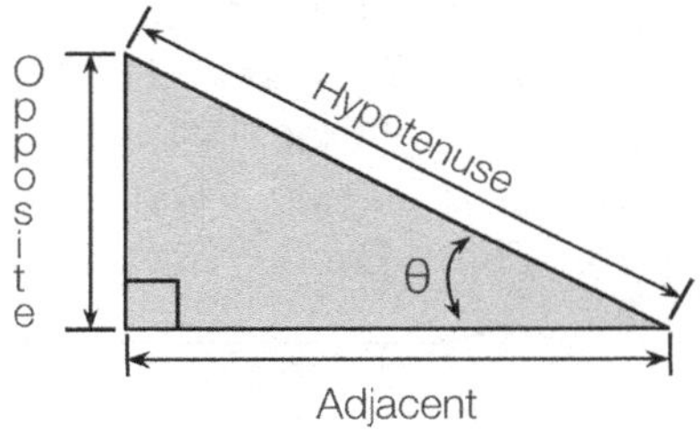

30 **Cosine** $\cos\theta = \dfrac{\text{Adjacent}}{\text{Hypotenuse}}$

31 **Tangent** $\tan\theta = \dfrac{\text{Opposite}}{\text{Adjacent}}$

Laws of Sines and Cosines for Any Triangle

32 **Sine Rule** for finding a missing angle when given an angle and two sides, or for finding a missing side when given two angles and one side.

$$\frac{a}{\sin A} = \frac{b}{\sin B} = \frac{c}{\sin C}$$

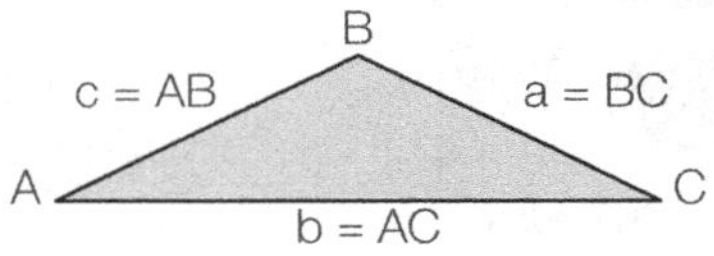

33 **Cosine Rule** for finding an angle measure when given all side lengths or finding a missing side when given the other sides and one angle measure.

$a^2=b^2 + c^2 - 2bc\cos A$, $b^2=a^2 + c^2 - 2ac\cos B$, $c^2=a^2 + b^2 - 2ab\cos C$,

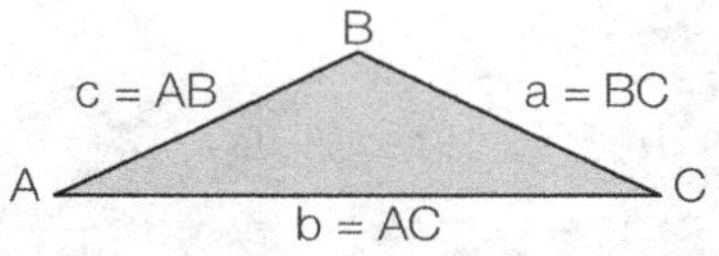

Notables (Fillable)

Ref: Page #: Line #:

Ref: Page #: Line #:

Ref: Page #: Line #:

Ref: Page #: Line #:

Notables (Fillable)

Ref: Page #: Line #:

Ref: Page #: Line #:

Ref: Page #: Line #:

Ref: Page #: Line #:

Index (Fillable)

Keyword, page#:ID#:line#

Index (Fillable)

Made in the USA
Monee, IL
08 November 2026

67946334R00070